This Book Belongs To:

_____

_____

"May the Lord bless you and keep you. May the Lord show you His kindness. May He have mercy on you."
                    Numbers 6:24-25
                    (International Children's Bible)

I dedicate this book to Almighty God who met me right where I was outside in my butterfly garden!

To my citizen scientists who supported and helped me with this book — John Jr., Joel, Keaton and Jade.

Thank you to editor Kristi Miller for her overwhelming love for Jesus Christ and for her guidance, advice and encouragement to finish this book; to John Thornton Jr. for the illustration inspiration; to Josie Renken for her truly beautiful talent in bringing the illustrations to life; to Raisa Gorielova for her creativity in the book design and to Mr. Nate who believed I could turn my experience into a book.

— T. T.

God's Fingerprint
The Lifecycle of a Monarch Butterfly
Copyright © 2021
ISBN: 978-1-7379653-1-2
Written by Tara Thompson
Edited by Kristi Miller
Illustrated by Josie Renken

Printed in the United States of America. All rights reserved.

All rights reserved solely by the author, Tara Thompson. The author guarantees all contents are original and do not infringer upon the legal right of any other person or work. No part of this book may be reproduced in any form without the permission of the author. The views expressed in this book are not necessarily those of the publisher.

This book is protected by the copyright laws of the United States of America. No portion of this book may be stored electronically, transmitted, copied, reproduced, or reprinted for commercial gain or profit without prior written permission from the author, Tara Thompson. Permission requests may be emailed to *godsfingerprintsinnature@gmail.com*. Only the use of short quotations for reviews or as reference material in other works is allowed without written permission.

May be purchased in bulk for educational, business, fund-raising, or sales promotional use. For information, please email *godsfingerprintsinnature@gmail.com*.

Scripture quotations are referenced and have been taken from:
International Children's Bible. @ 1986, 1988, 1999 by Thomas Nelson, Inc.
New International Version, Adventure Bible. @ 2103 by Zonderkidz.
The Passion Translation Bible. @ 2020 by Passion & Fire Ministries, Inc.

God's Fingerprints
The Lifecycle of the Monarch Butterfly

Written by Tara Thompson
Edited by Kristi Miller
Illustrated by Josie Renken

The monarch butterfly is one of God's beautiful creatures found in nature. Each monarch butterfly is wonderfully and perfectly made by God but also has a special job to do.

The monarch butterfly flies from flower-to-flower sipping nectar.

When it stops to sip, some of the pollen from the flower sticks to its feet. As the monarch butterfly travels to the next flower, the pollen from its feet transfers to that flower. This process is called pollination. Pollination helps fruits, vegetables, and flowers produce new seeds.

Just as a butterfly spreads seeds by pollination, we also can spread seeds to plant and grow flowers and vegetables in our gardens. Did you know we can plant other seeds too?

When we talk to others about our love for Jesus Christ, we call this planting a seed. The seed we plant is the message of our salvation through Jesus Christ. Salvation is the gift God gives every person who loves and accepts Jesus Christ into their heart.

This gift is to us all. It is not something we can earn by being good. God freely gives us this gift because He loves us so much that He has endless grace and mercy for us. Grace allows God to forgive us over and over again. Mercy is receiving God's forgiveness even though we don't deserve it.

The more we plant the seed of Jesus Christ in others, the more our love, trust, faith, and relationship grows with Him.

> "But what is the seed that fell on the good ground? That seed is like the person who hears the teaching and understands it. That person grows and produces fruit, sometimes 100 times more, sometimes 60 times more, and sometimes 30 times more."
> Matthew 13:23 (International Children's Bible)

Before a monarch butterfly has its beautiful wings, it begins life as a monarch caterpillar.

Monarch butterflies lay their eggs on only one type of plant, the milkweed. Milkweed is the only thing monarch caterpillars eat. They cannot survive without milkweed.

Monarch caterpillars devour the milkweed. Devour means to eat very hungrily and quickly. A monarch caterpillar can devour an entire milkweed leaf in less than five minutes and an entire plant in two weeks! And that's just one caterpillar!

The same way monarch caterpillars devour their one and only food source, milkweed, we also need to devour God's word and promises in the Holy Bible.

Like the monarch caterpillar, the hungrier and quicker we devour God's word, the faster we grow in our faith.

> "So until I come, be diligent [dedicated] in devouring the word of God, be faithful in prayer, and in teaching the believers."
> 1 Timothy 4:13 (The Passion Translation)

As the caterpillar continues to eat the milkweed and grow, it quickly becomes too large for its skin. When this happens the skin begins to molt, which means the skin sheds off and reveals a new layer of skin.

The new skin is soft at first, which provides very little protection but soon the new skin hardens and molds itself to the caterpillar.

The caterpillar molts or sheds its skin five times as it continues to eat the milkweed and grow. The time between when a caterpillar molts is called an instar.

Just like a caterpillar that grows and sheds its skin as it devours milkweed, when we read and devour our Bible, we grow in our relationship with Jesus Christ and shed things too.

We learn what pleases God so that we stop doing behaviors that are not pleasing to Him. Some of the behaviors we may shed off as we grow could be arguing with our brothers and sisters, lying, or not listening to our parents.

> "But you were taught to be made new in your hearts. You were taught to become a new person. That new person is made to be like God — made to be truly good and holy.
> Ephesians 4:23-24
> (International Children's Bible)

The monarch caterpillar continues to devour the milkweed until it eats every leaf on the plant. This can take around 10-14 days.

After this time the caterpillar leaves the milkweed and travels up to 20 feet to find its next resting place. Once there, it hangs upside down and forms the shape of a "J."

The caterpillar in the shape of a "J" is completely vulnerable to nature and its elements but the caterpillar does not think about any of these dangers. It does not worry about other creatures or storms. Instead, it focuses on what it is meant to do next – form a chrysalis.

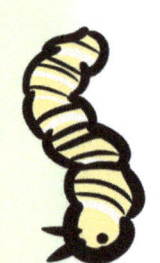

The chrysalis is the outer shell that protects the caterpillar as it transforms into a butterfly.

For the caterpillar to form its chrysalis it must fully surrender to God's plan for its life. To surrender means to obey or submit. The caterpillar must obey God's plan to complete its purpose on earth.

Similar to the monarch caterpillar, when we are filled with God's word and submit fully to it and obey, we become filled with faith. Faith is what helps us no matter what happens in our life. Even when a storm comes that turns our life upside down, we have faith that we are securely anchored in our "J" — Jesus Christ. An anchor is what holds a ship steady in a storm. Jesus Christ is the one who holds us steady through all things.

> "We have this certain hope like a strong, unbreakable anchor holding our souls to God himself..."
> Hebrews 6:19 (The Passion Translation)

Did you know that God plants gifts into everything He creates? That's right. Even you and I have special gifts God planted in us. The monarch caterpillar does too!

The chrysalis is the special gift God planted inside the caterpillar. The chrysalis stays hidden inside of the caterpillar waiting to be revealed until after it molts or sheds its skin for the last time.

Just like a monarch caterpillar's chrysalis is inside where it cannot be seen, the Holy Spirit is inside of us and cannot be seen, unless we choose to show Him.

As we grow in our relationship with God and Jesus Christ, we are protected and transformed by the Holy Spirit inside of us. The Holy Spirit speaks to us and guides us in the way we should live our lives pleasing to God. He gives us the power to change our old behaviors into new behaviors that glorify God, His kingdom, and His love and mercy. The Holy Spirit also enables us to stand bold in our beliefs and defend our family and friends.

> "For God's promise of the Holy Spirit is for you and your families, for those yet to be born and for everyone whom the Lord our God calls to himself."
>
> Acts 2:39 (The Passion Translation)

The chrysalis first appears light-green in color. As the days pass, the light-green chrysalis darkens to a blackish color, and eventually, it becomes completely transparent. Transparent means you can see what is inside of something. When the chrysalis is transparent the butterfly's wings can be seen!

Inside the chrysalis, God takes all the parts of the caterpillar and changes it into a completely new creature – a Monarch Butterfly.

Just like God transforms the caterpillar into a butterfly, when we read our Bible to learn more about Jesus Christ, God takes all of our sins away and changes us into a completely new creature — a Christ-like child of God.

> "If anyone belongs to Christ, then he is made new. The old things have gone; everything is made new!"
> 2 Corinthians 5:17
> (International Children's Bible)

After the monarch butterfly emerges from its chrysalis, it hangs upside down for up to two hours. This time allows its wings to stretch, dry, and harden in the sun and heat.

When the wings are dry, sturdy, and firm, the butterfly is ready to spread its wings and fly. The process of a butterfly emerging from its chrysalis is called eclosion.

Jesus Christ wants us to be firm in our faith just like a butterfly's wings need to be firm to fly. God wants us on fire for Him, not cold, not warm, but **HOT** so we can spread our wings and fly too.

We stay on fire for God by always worshipping Him, praying, and reading His word in the Bible. To worship God means we show Him how much we love and respect Him by always including Him into every part of our life.

When we speak God's word, pray to Him, study the Bible, and focus our minds on what He is telling us, we bring more than just words into our life; we bring **POWER** into our life! We have the same power and authority as Jesus Christ when we believe and speak His word in faith.

> "The kingdom of God is not a matter of talk. It is a matter of power."
> 1 Corinthians 4:20 (Adventure Bible)

Once the monarch's wings are firm and steady, it takes flight. The monarch butterfly spends the rest of its life traveling from flower to flower sipping nectar in search of milkweed plants to lay its eggs.

The monarch butterfly knows the milkweed is the one and only source of food its baby caterpillars need to grow and survive. The monarch butterfly individually lays each egg on the milkweed. One butterfly can lay up to 400 eggs!

Just as the butterfly is commanded to spread its wings and take flight, we are commanded to take flight by sharing the love of Jesus Christ with others.

When we share Jesus' love with others, we become His disciples. A disciple is a person who believes in Jesus Christ and follows His ways.

> "Then Jesus came to them and said, "All power in heaven and on earth is given to me. So go and make followers of all people in the world. Baptize them in the name of the Father and the Son and the Holy Spirit. Teach them to obey everything that I have told you. You can be sure that I will be with you always. I will continue with you until the end of the world."
>
> Matthew 28: 18-20
> (International Children's Bible)

There are approximately 17,500 species of butterflies. Each butterfly is beautiful in its own right. Did you know that no two butterflies are exactly alike?

Everything God creates is unique, which means everything He creates is one of a kind. God also created each one of us unique in His image.

Every person is unlike anyone else on this earth. He made each one of us unique because each one of us has a unique purpose to fill on this earth that no one else can.

Have you ever thought about all the parts of your body? You have legs, feet, arms, hands, eyes, and so many other parts! Now put all of those parts together and you have a whole body. When smaller parts come together to make a whole, we call it united.

Each person who accepts Jesus Christ into their heart is immediately united with other believers. When all believers of Jesus come together, we become one whole body of Jesus Christ. It is only because of Jesus' love and salvation that we are united together as one body.

> "God has chosen you and made you His holy people. He loves you. So always do these things: Show mercy to others; be kind, humble, gentle, and patient. Do not be angry with each other, but forgive each other. If someone does wrong to you, then forgive him. Forgive each other because the Lord forgave you. Do all these things; but most important, love each other. Love is what holds you all together in perfect unity."
> Colossians 3:12-14 (International Children's Bible)

Prayer of Salvation

Dear Lord Jesus,
I know that you made me unique and want me to trust, love, and obey you with all my heart.
I know I have sinned by doing things my way and not your way and I am sorry.
I ask for your forgiveness.
I believe you died for my sins and rose from the dead.
I want you to come into my heart and life.
I turn from my ways and sins.
I want to trust and follow You as my Lord and Savior.
Save me now and forever.
In Your name, I pray.
Amen.

More Interesting Facts About Butterflies

The top speed of a butterfly is 12 miles per hour. *(12 tribes in the Bible.)*

The wings of some butterflies are marked with patterns that look like letters of the alphabet, as well as, numbers. *(God speaks to us with letters and numbers.)*

A butterfly cannot fly if its body temperature is below 86 degrees. *(God wants us to remain HOT for Him.)*

Butterflies do not fly at night or in the rain. Only in the light. *(We are called to be the light of the world.)*

Butterfly wings are transparent and covered by thousands of scales. The colors we see are reflections of various colors through their scales. *(We can't see God but we know He's there. God is light radiating and illuminating a bright light with many beautiful colors.)*

Butterflies have a liquid diet. *(Jesus said those that come to Him will never be thirsty.)*

When a butterfly emerges from the chrysalis it has to put its mouthpiece together. It's in two pieces and it must work them together to survive. *(We trust the word of God by hearing the word, someone must be speaking for us to hear it and we are commanded to go speak of Jesus Christ to others — we must use our mouth!)*

The early colonists of North America thought the gold rim around the monarch chrysalis resembled a king's crown so they named the butterfly "Monarch." *(Jesus Christ is the King of Kings.)*

How to Start Your Own Butterfly Garden

The number of butterflies in nature is declining, which means each year there are fewer and fewer butterflies. We need to do what we can to help God's creatures survive.

The first thing we can do is teach others about butterflies. We can also make a butterfly garden!

A butterfly garden ensures butterflies have a place to lay their eggs and the best part — you get to watch them grow as you observe each phase of the cycle you just learned about!

When starting a butterfly garden, it is important to remember a few things:

1. Always remember your plants will be used by butterflies to sip nectar and caterpillars to eat so we never want to use pesticides on our plants because this can hurt the caterpillars and butterflies.

2. Butterflies need to have warmth to fly so when you look for your location to plant your garden be sure there is plenty of sunlight.

3. Butterflies and caterpillars also need protection from wind and predators which can be done by planting your plants near a tree, fence, or house, to block the wind, just make sure the plants get sun at least 6 hours a day. If you plant your plants in pots then you can move them out of the wind.

4. As we learned in our book if you want to attract Monarch butterflies then you need to plant Milkweed. Monarch caterpillars **LOVE** to eat so be prepared with extra milkweed!

5. There are many other types of plants that will attract multiple species of butterflies. You can find a few websites below to help you find plants specific to where you live.

For additional resources visit:
Monarch Butterfly Garden at *monarchbutterflygarden.net*
National Garden Bureau at *ngb.org*
The Butterfly Website at *butterflywebsite.com*
Monarch Watch at *monarchwatch.org*

I would love to see your butterfly gardens! Ask your parents to email me your pictures at *godsfingerprintsinnature@gmail.com*

www.ingramcontent.com/pod-product-compliance
Lightning Source LLC
Chambersburg PA
CBHW051307110526
44589CB00025B/2963